STUMP FARMS AND BROADAXES

STUMP FARMS AND BROADAXES

by JACK MOULD

hancock
house

HANCOCK HOUSE

ISBN 0-919654-54-1

Canadian Cataloguing in Publication Data

Mould, Jack, 1909-
Stumpfarms and broadaxes

ISBN 0-919654-54-1

1. Frontier and pioneer life—British Columbia—Bulkley Valley. 2. Bulkley Valley, B.C.—Hist. 3. Burns Lake district, B.C.—Hist. I. Title.

FC3845.B78M6 971.1'32 C76-016018-X
F1089.B78M6

PRINTED IN CANADA

Published by:

hancock house

HANCOCK HOUSE PUBLISHERS LTD.
3215 Island View Rd.
Saanichton, British Columbia V0S 1M0
12008 1st Ave. S.
Seattle, Washington 98168

Dedication

To my wife, Edna, who always believed this book would be published.

Table of Contents

Chapter

Introduction

The Bulkley Valley-Lakes District area generally followed the national pattern of settlement and development, but some facets of its early growth were unique.

No early explorers travelled this area. At the middle of the 19th century it was known to white men only as a vague area North of the Cariboo gold fields and South of the Babine fur trading route. It did not even have a name until 1866, when the ill fated Collins Telegraph line was built through B.C. to the confluence of the Skeena and Bulkley Rivers. The line locators were the first explorers in the area and the River was named at that time for Colonel Bulkley, who was in charge of line construction.

It was as recent as the turn of the present century that the rebuilding of the abandoned telegraph line, the development of river transportation inland to Hazelton, and talk of a railway to be built provided the incentive that brought in the pioneers. They came by foot and packhorse, and established a foothold in what was a beautiful untamed wilderness only three quarters of a century ago.

The railway was completed in 1914, but as the outbreak of World War One also occurred that year there was no settler immigration of note until after that affair was resolved. The first influx of settlers, via the railway route, commenced in 1919 and continued throughout the 20's until the Depression brought immigration to a temporary halt. These post War One immigrants were destined to become a forgotten phenomenon—the stump farmers—broadaxemen of the 20's and 30's. They were the first, and were also to be the last of the homesteaders. Many of them, still living today, witnessed the introduction of the hewn tie industry into the area, watched it develop until it became an entire way of life for two decades, then disappear so completely that today the tools that were part of it are museum pieces.

How many people living here today can identify with a broadaxe; snaabsaw; picaroon; roughlock; scoring axe; or dryass? Thirty years ago these were household words throughout the Bulkley Valley and Lakes District. Hewn tie making was not restricted to this area. In fact, as a matter of expedience, most of the ties which went into the initial railway construction were produced from hemlock stands in the Terrace-Usk areas. However, pine is a generally preferable species for railway ties and when the Post War One demand for replacement and export requirements occurred, most of the production came from the lodgepole stands further inland.

Ties were produced in varying amounts along the entire railway from Prince Rupert to Red Pass Junction and the Stumpfarm-Broadaxe life style described in the following chapters was generally applicable throughout.

The Bulkley Valley-Lakes District locale has been chosen for this glance into the past for the following reasons:

1. Authentic "cut" statistics applicable to the area from Prince Rupert to Fraser Lake, were available from the Forest Service Records.
2. The "Summit" area, (Houston to Burns Lake) produced more hewn ties than any comparable section along the entire railway and was also typical "stump farm" country.
3. The writer was raised on a "stump farm" at Rose Lake on the boundary of the Bulkley-Valley and the Lakes District and started hewing ties in 1923 at the age of 14 so has first hand knowledge of the early years in this area.

So welcome to a trip into our history with one who was a part of it.

Jack Mould

1

The Settlers

The post-war One settlers were attracted by the opportunity to obtain free land. The procedure was simple. “File” on 160 acres of Crown land for a fee of $10.00, reside on it for at least 6 months per year for three years, do a specified amount of improvements, and you received your Crown Grant. If you were a returned soldier, the government would supply the land and also a small loan for the purchase of equipment.

However, there was a slight catch. Most of the arable land was already alienated, held either by the pioneer ranchers or speculators, who like the poor, seem always to have been with us. The available land was away from the valley floor, difficult of access, and heavily timbered and was generally unfit for farming, but the settlers were here, had burned their bridges behind them and so had to dig in and make the best of it.

Some came from the prairies. They travelled with their effects and stock in boxcars, sometimes several families banding together to share travelling costs. Their first priority was to erect some kind of shelter from the material available, and some of these first structures were rough and primitive. Next was to rustle stock feed from sloughs and wild meadows but in most cases the first winter was longer and more severe than anticipated and heavy stock losses were suffered. The immigrants' dreams of becoming self sustaining soon vanished but they held on grimly, living in part from trapping, road making and working in the woods.

There was a fair percentage of Scandinavian

immigrants, mostly bushmen and experienced axe-men. They found themselves in a familiar environment and quickly got involved in the developing hewn tie industry. These people had little trouble adapting to their new homeland and their descendants hold many positions of importance in the country today.

This was also the era of "the Green Englishman". A good proportion of the new settlers were English, coming for the most part from cities and industrial backgrounds. They were totally unprepared for frontier life, and were further handicapped by their honest belief that they were superior, by right of birth, to their Colonial neighbours. Canada had no national status then; it was merely another colony—there were no Canadian citizens—only "British subjects". The newly arriving English seemed to feel

they were in the position of visiting poor relations, and were here only as a means of making a fortune, and then going 'ome. (Very few of them did either). They were hard working but seemed to feel it was below their dignity to accept advice from aliens and "colonials", and chose to learn for themselves via the long hard way route. During this period the Green Englishman jokes flourished.

"Which way is that tree you are cutting going to fall? 'Ow do I know. I'm not a bleedin' prophet." One scared Englishman to another, "Did you 'ear the owl?" "Yus, I did, but what's that 'owlin'?"

"Were you in the boat when the boat tipped ovah?"

"No, I was in the bloody water!"

The Englishman lost in the woods is found almost dead from thirst. His puzzled rescuers inquire if he hadn't encountered any creeks. The reply—"Oh yes, several of them, but dash it all I never had a cup", and on and on they went.

The English however are a hardy race. They weathered the elements and the jokes, and eventually became good Canadians. It just took them a little longer than the others.

A smattering of other nationalities made up the roster of first settlers—the stump farmer—broadaxemen of the forgotten decades.

Smithers area, 1920. Much of the better arable land of the lower Bulkley Valley was found here but was already alieniated when the post World War One immigrants arrived. [Photo courtesy of Mrs. W. H. Morris]

Telkwa, early 20's. Log houses were predominant during this early period. [Photo courtesy of Mrs. W. H. Morris]

Rose Lake, 1922. The author's parents, Jim & Nellie Mould. He had served as a British soldier in the Boer War [Africa] then India, finally France World War One. She was the mother of 9 children and had never travelled 50 miles from her birthplace in beautiful Winchester, England until making the epic 21 day trip to North Central B.C.

Forestdale, 1925. Some settlers found work with railway maintenance crews. The starting wage was 25¢ per hour, and the jobs were mostly only temporary. [Photo courtesy of Mrs. E. F. Rowland]

2

Olaf Hanson (1882-1952)

Olaf Hanson was born in Sweden and immigrated to Alberta in 1904. Here he became involved in the production of the hewn ties required for the construction of the Grand Trunk (later CN) Railway, then slowly being pushed westward.

His ability in the timber industry became quickly apparent and when the decision was made to start construction from the planned western terminus of Prince Rupert, he obtained the contract to supply the ties and timbers necessary to construct the railway to its meeting place with the builders working from the east—this was destined to be at Fort Fraser in 1914.

The Hanson Lumber and Timber Co. was formed about 1906 and its first office established at Prince

Topley, [Circa 1924]. A packtrain leaving for Topley Landing, Babine Lake. It carried supplies for diamond drillers doing initial exploration at "Copper Island", now the site of the active Granisle Mine. The packtrains brought back the drill cores for rail shipment to assay facilities. The round trip took three long, tough days. Today the 30 mile trip to Granisle can be made over a paved highway in 30 minutes. [Photo courtesy S. T. Strimbold].

Rupert. Broadaxemen were recruited, mostly from Sweden and Eastern Canada to hew the ties, main camps being established in the Terrace and Usk areas. These operations were not the start of the hewn tie industry, per se, but a part (an important one) of the railway building project. After its completion, the Hanson Co. turned to the production and export of cedar poles which were cut adjacent to the Skeena River, then floated down to the rail loading points at Cedarvale and Skeena Crossing (Nash "*Y*").

The end of the war in 1918 heralded the end of a period of stagnation in the Central B.C. area. By 1919 settlers began to arrive, and coincidently a new demand for hewn ties developed. Production jumped from 50,000 in 1918 to 280,000 in 1919, and the stump farm—broadaxe era was born.

The Hanson Co. immediately entered into this new

tie production field, establishing camps in the high quality lodgepole stands of the upper Bulkley and Burns Lake areas. Of more significance to the ultimate development of the area, they initiated a policy of issuing subcontracts to the fledgling settlers for the making and delivery of small amounts of ties to their nearest rail sidings. Other large operators entered the hewn tie production field at this time and some were later to sub-let contracts for small amounts, but the Hanson Co. pioneered in this field. Their policy of advancing some payment as the ties were delivered to the sidings proved a boon to many settler families during the first years of stump farming.

In 1925 the Hanson Co. moved their administration facilities from Prince Rupert to Smithers; (a small branch office had served prior to that time). The new building, from which they were to operate for the following 27 years was the most impressive office complex built in the area to that date, and it was also to be the home of the Royal Bank branch for a considerable period of time. It is currently occupied by Apollo Auto Parts Ltd., its connection with the past forgotten.

Olaf Hanson was a remarkable character. In addition to being one of the last self made timber barons, he was a notable politician. He was B.C.'s Swedish Consul for 32 years and that commitment kept him in the South much of the time, his extensive business in this area being handled by an ultra efficient staff. However, he did establish an elaborate "summer home" at Lake Kathlyn, where he and his family spent most of the summers.

He was elected liberal M.P. for Skeena and

represented the riding at Ottawa for 15 years. He was not a good public speaker, as he retained a strong Swedish accent so he usually had more fluent orators carry the brunt of his electioneering. On one occasion, however, when running for his third term, he did appear on the podium, and startled his audience by announcing in a ringing voice "I've been s(h)itting down at Ottawa for the last two terms, and I full expect to go back this Fall for the t'ird!"

Olaf Hanson died at Vancouver in 1952. His company was liquidated at that time, and its contribution to the development and history of the country soon forgotten, except by a few old timers who occasionally remember the lost decades of another era when they stump farmed and "took out ties for Hanson".

Circa 1910. Moving ties up to the "end of steel" for track laying. Ties for construction of Grand Trunk Pacific Railway were supplied by Hanson Lumber & Timber Company. [Photo courtesy of Smithers Museum].

Olaf Hanson [right] and Austen Goodenough, the Field Superintendent and driving force of the Hanson Lumber & Timber Co., pose by a deck of valuable cedar poles. Date unknown. [Photo courtesy of Smithers Museum].

Early 20's, Decker Lake area. A newly completed "little grey home in the West". Note rough door and latch assembly and general lack of frills. Axework is inspiring feature. [Photo courtesy of R. Buer]

Topley [Graveyard Lake area], 1920. The Strimbolds visit beloved and respected Ole Fjästed. Mr. Fjästed was a very gifted musician. In this little log cabin he taught violin and piano music to many children. [Photo courtesy of Mrs. E.G. Strimbold]

Smithers area [Driftwood], circa 1925. Oscar Lundstrom. His Scandinavian heritage is reflected in the skilful axework pictured on this newly built barn. He was very active in the hewn-tie industry. [Photo courtesy of Mabel Riley [Nee Lundstrom].

Looking westward from the summit to the east end of Bulkley Lake. This was the heart of stump-farm and broadaxe country.

This is what it was all about. A beautiful stand of lodge-pole pine, the ideal size for tie making.

3

The S. Anderson Company

Sivert "Bull River Slim" Anderson arrived in Saskatchewan from his native Norway in 1913, when he was about 15 years old. A few years later he moved further west, and by 1918 was engaged in hewing railway ties in the Bull River area of south-eastern B.C. (and picked up his nickname there). His next move was to Engen on the C.N.R. in north central B.C. where he again engaged in the hewn tie industry, but was unable to get permanently established.

In the spring of 1922 he arrived in Burns Lake, still seeking his place in the sun, (and in the tie industry). He asked many questions, walked many miles, and eventually located tie timber to his liking 3 miles south of Rose Lake and 16 miles west of Burns Lake.

Behind the Anderson Co. ram pasture, an unknown tiemaker has "drank his fill" and dreams blissfully of taller trees and easier hewing.

1923. Sivert Anderson. He made a valuable contribution to the original development of upper Bulkley Valley and Lakes District areas. [Photo courtesy of E. F. Rowland]

1922, Burns Lake area. Sigurd Esplenen of the S. Anderson Co. offers Fokke Anderson a "chew of Snoose". This concoction was the "pot" of the era and was indulged in more than tobacco. "Snoose" was not actually chewed. It was packed loosely between the lower lip and front teeth and just laid there. [Photo courtesy of Mrs. Fokke Anderson].

There was a log cabin and barn, recently abandoned by a discouraged stump farmer, available for use as a ready made camp, and Bull River Slim realized this was the opportunity he had been long seeking. He interested three other young Scandinavian immigrants and with them formed the S. Anderson Company.

They obtained cutting rights to the timber, and that fall started their first operation. They were ambitious, experienced woodsmen and they put out 27,000 ties before spring, doing most of the work themselves. Another successful season followed and when they moved to Decker Lake in 1924, the S. Anderson Company was ready to establish itself as the leading tie contractor of the upper Bulkley Valley and Burns Lake area.

1922. Mrs. Charlie Mould [A World War One English war bride] on their Soldier Settlement place south of Rose Lake.

Previously the sub-contracting of tie contracts had been a minor part of the industry, handled almost exclusively by the Hanson Lumber Co. of Smithers, but the S. Anderson Co., its credibility established, was successful in obtaining large volume contracts from the railway, and took over most of the increasing subcontracting business in the area from Houston east to Endako.

The stands required to support the original large scale operations were becoming depleted, but there was still plenty of timber available in amounts suitable for two to six man and settler operations and these soon became prevalent throughout the area. These did not produce large enough amounts to warrant dealing directly with the railway, who issued contracts for only large volumes and did no financing, so the new crop of operators marketed their products via the sub-contracting method.

The contractors were not philanthropists. They charged 10¢ per tie (from the base rate of 60 - 70¢) for their services, which involved advancing part payment as ties reached the landings, paying Forest Service charges, and the loaders' wages, all of which were debited to the subcontractor when final settlement was made after the ties were shipped. The S. Anderson Co. charged the same commission as their predecessors but gave better service, establishing a general store at Decker Lake from which subcontractors were supplied with groceries, equipment, horse feed, etc. The Company also met

Topley, late 1930's. Everybody worked. Archie Strimbold and brother Frank hit the woodpile. Archie now runs a mechanized logging show. This picture is proof that he "paid his dues."

1919, Decker Lake. The Fred Sauls came from Victoria and were among the first of the stump farmers. They lived in this army tent while making their initial clearing. Two later generations now have homes on this now beautiful property which overlooks Decker Lake and is appropriately named "Saulhaven". [Photo courtesy of L. G. Saul].

payrolls as the ties were being produced and in some cases grubstaked people during the off season. In addition, Bull River Slim was constantly in touch with the operators moving from camp to camp giving advice and assistance.

The "cut and get out" phase of the industry was now drawing to its close but there was no drop in production, as the numbers of small contractors

Modern "Saulhaven" today.

increased. The seven years from 1924 to 1930 were the most prolific of the stumpfarm—broadaxe era, over 5,000,000 ties being shipped out during that period from the Bulkley Valley—Lakes District areas, and the industry became the most important factor of the slowly developing economy.

The stump farmers gradually became integrated into the tie industry as they realized the futility of trying to clear and farm marginal land. They revised their life style, now being content to use their homesteads to raise gardens, and grow enough hay to feed a milk cow or two and a team of horses, but basically as a home base for their woods operations. There is really nothing new about the contemporary crop of "hobby farms"—many of the areas so in use today served the same purpose forty years ago, before the Depression and resulting collapse of the hewn tie industry forced their abandonment.

Some of the single professional tie makers who had previously moved into the large camps in the fall, and left the area for parts unknown in the spring, also became subcontractors at this time. A number of them located on the south side of Decker Lake and other attractive locations where they remained during the off seasons, enjoying a free life—raising small gardens, making home brew, and fishing, in addition to locating timber, cutting roads and generally preparing for the following season's operation. They did not bother with livestock, usually hiring the nearest stump farmer to do their hauling.

There were still some of the ever restless "floater" types drifting in and out of the district. For these, the Burns Lake beer parlor served as an unoffical hiring hall, and for their benefit the S. Anderson Co. provided a unique facility—the famous (or infamous) Decker Lake "Ram Pasture".

The Ram Pasture was a building on the hill behind the store, always open and available to all and sundry as a "free of charge" stopping place. It housed a stove, a few cots and a minimum of other furnishings. No provision was made for maintenance, the users occasionally swamping it out as their consciences dictated. Over the years it was the scene of some awesome binges, especially during spring breakup periods, but it generally served a useful purpose and many weary and/or hungover bushwackers found it an ideal place in which to recuperate or just mark time.

The S. Anderson Company played an active role in the early development of the Bulkley Valley-Lakes District area throughout all the stumpfarmer—broadaxe years, and far beyond into the small sawmill era which followed. It has been gone from the rapidly changing scene for many years now, but the man who personified the S. Anderson Company, now nostalgicly referred to as "Old Bull River", still lives in retirement in the Burns Lake country.

4

The Stump Farmers

Many of the first settlers had very little knowledge of what they were getting into when they came to central B.C. Only a few were to acquire pre-emptions with farming potential, as most of the unalienated land, being at the higher levels, was heavily timbered, susceptible to summer frosts, and difficult of access. The cutting of miles of wagon roads into their homesteads was generally the first formidable task faced by men, who in some cases had never previously handled an axe or saw.

No, Virginia, there were no bulldozers, no roads, no cars, televisions, telephones, radios, and at first, no schools. In fact, nothing but a double handful of inexperienced but optimistic people with their hopes and dreams and a few farm animals, scattered over a

1928. There were two sure crops in the 1920's, Kids and stumps: Pictured are those of the Axel Andersons, immigrants from Sweden

vast, beautiful and often hostile land. A passenger train which ran three times a week, on a very uncertain schedule, was the only link with the outside world—and oh yes—the stumps.

Blissfully unaware that in most cases the land was unsuitable for growing anything but trees, the settlers literally worked like beavers to clear it. The cutting, piling and burning was back breaking work in which the entire family shared. It was a hard task but it could be done. Getting rid of the stumps was another matter. Some could be grubbed (what a discouraging word), and some pulled out by horses, if and when acquired, but the larger ones could only be

1922, Rose Lake. Len Ailport and stepdaughter Edna, on his stump farm in the Summit area. In the background is timber suitable for hewn-tie production, a good combination in the 1920s.

moved by stumping powder, which cost money, a commodity usually not available. So the stumps remained, their number increasing year by year as the tree cutting continued. In some cases, the stumps were the only crop of any extent raised which prompted the coining of the descriptive term "stump farmers", which was soon integrated into the jargon of the times.

It soon became obvious to even the most optimistic that a living could not be made raising stumps, and some gave up their dreams and either left the country, or competed for the few jobs available such as railway maintenance, etc. Others, reluctant to leave the homesteads they had travelled so far to

Rose Lake, 1922, Mrs. Nellie Mould. An English lady a long, long way from 'ome.

acquire, looked for part time work that would bring in some income, and still allow them to work at developing the land. The answer was to cut and sell railroad ties. Most had suitable timber on their homesteads and soon mastered the use of the broadaxe. Making ties was seasonal winter work as the Railway would only purchase ties cut between September and May when the sap was down, and in any case, they could only be hauled when snow was available for sleighing. This created an ideal situation for the stump farmers.

They did not sell their ties directly to the railway but took sub-contracts to deliver small amounts from main contractors, who advanced payment as the ties were delivered to the sidings. This allowed the stump farmers to live through the winter and when the final settlement was made after the ties were loaded out in the spring, there was usually a little balance due, enough to keep them going through the following summer.

After the first lean years most stump farms did

1926, near Topley. These ties are being hand skidded down to piles on the haul road by Walter Holmes and Bob Jannack. [Photo courtesy Ivan Holmes]

1926, Decker Lake. Henry Piche, son Paul and daughter Ruth [Mrs. E. F. Mould] pose in the "tie bush". The day's cut has been dragged out and piled on the haul road. [Photo courtesy Mrs. E. F. Mould]

produce gardens, enough feed for a team of horses, and a milk cow or two, and this combined with the tie income provided a reasonable living and the independence that those first settlers sacrificed so much to attain.

The stump farm-broadaxe way of life flourished through the 1920's and 1930's. It survived the Depression years (when the demand for ties dropped drastically), but it gave way to the machine age which reached this area in the 1940's. The machine and stump farmer-broadaxe cultures had nothing in common and since the machine age was definitely here to stay, the stump farmers and broadaxe men became a doomed species and were soon extinct, but they had done the hard work necessary to make it possible for lesser men to follow, and earned a respected place in the history of North Central B.C.

5

The Camps and Tiemakers

Producing hewn ties for an almost unlimited market was an attractive venture for those with know-how and available financing. Suitable Jackpine (Pl.) timber with a 11-15" diameter range, located within five miles of a railway siding was the basic requirement, and in the first few years there seemed to be an unlimited supply available on Crown land. Cutting rights were obtainable with a minimum of red tape.

Road location was important as hauling by horses and sleighs was only possible during the winter months and all adverse grades had to be avoided. The roads usually followed water courses down to the valley floor and on to the nearest siding, and as there was no surfacing involved, the cost of construction was minimal.

The camps consisted of a cook house, bunkhouse and horse barn built from logs obtained on the area. The buildings were chinked with moss, and very little lumber was used in their construction. A couple of good axemen, with a horse to skid out the trees, could cut the logs and erect the crude structures in a matter of a few days. The tie cutting commenced in September, and continued as long as there was snow available for hauling, usually around mid April.

The larger camp rosters varied from 10 to 20 men with hauling teams in proportion. The tie makers worked on a "piece work" basis and the average production for a season was between 2 and 3

1923. An exception to the rule—a neat tie camp building, also in the Topley area. The wide plank nailed between two trees is a "saw filing rack", a necessary installation at the tie camp. [Photo courtesy of Ivan Holmes]

1928, Rose Lake. Rather than peel their ties during the cold winter months, some operators hauled them to the landings and had them peeled there in the spring. Nine year old Jimmie McKeand is not playing games. He is peeling ties for 4¢ per tie. [Photo courtesy Mrs. E. F. Rowland]

thousand ties per man. They worked alone in areas or "strips" allocated to them, first felling the trees, which each contained from three to six tie lengths, then limbing and measuring, "scoring" and hewing down to 6 - 8" thickness. When sawn into eight foot lengths the trees became ties, and all that remained was to pile them on the roadside ready for hauling. The men received 20 cents per tie, and were charged $1.00 per day for board. The average cut per day per man was 20 to 25 ties per day, the suitability of the available timber being the governing factor.

Over the years, fabulous stories of individual production were related from time to time but like

1920, Burns Lake area. An artist at work "creating" hewn ties. This axeman is the late Folke Anderson, shown working in an excellent stand of hewn-tie timber. [Photo courtesy of Mrs. Folke Anderson]

most "Paul Bunyan" type stories they were hard to substantiate and grew with each telling.

I knew a man who averaged a daily cut of 40 ties throughout a season, and another who hewed 100 ties in a single day. The latter had others doing the falling, limbing and measuring. These were mild feats in comparison with the "Paul Bunyan" type legends (some of which still survive), but nevertheless awe-inspiring when considered on their own merits.

The 40 a day man would fall, by crosscut saw, about 12 trees (12-16" diameter), limb and measure them, then "score" and hew off the sides—640 feet. This would involve swinging his five lb. scoring axe about 10,000 times and his 7 or 8 lb. broadaxe 6,000 times for a total axe lift of 95,000 lbs. A sportswriter commentating on the effort expended in the sport of curling once pointed out that in a ten end contest, each curler lifts and throws a 40 lb. rock twenty times for a total lift and throw of 800 lbs. The 40 a day tiemaker lifted and swung as much weight every day as a curler does in more then 100 contests. To round out his days work, the tie man also had to pull or carry his 40 ties to the roadside piles... 40 ties averaging 200 lbs.—4 tons.

The man who hewed a 100 ties in a day—well, it rather staggers the imagination. 1600 feet, well over a quarter of a mile of hewing the sides of trees down to a 6 to 10" face, involving the lifting and swinging of over 200,000 lbs of axe weight and every blow had to be accurate.

The "professionals" who worked in these camps were a distinctive breed. Of all nationalities, they had in common their strength and skills and engaged in this arduous work for the feeling of creativity and freedom they derived while making their living this way.

I suppose the attraction was mostly the "piece work" something that is generally despised in today's over-unionized society.

Although the tie maker worked harder for much less than his modern counterpart in the logging industry, he was to a greater extent his own man. He was paid according to his production and he owed allegiance to no Company or union. If he did not like the camp boss, the timber allocated to him, or just had the urge to move on, he just collected his due, shouldered his blankets and tools and left. During the 20's his services were always in demand, and although those that moved often were labelled "short Stakers", they were always welcome in the next camp as long as they produced.

The large camp operations only flourished for a few years as the compact bodies of suitable timber were soon depleted. However the industry continued to thrive as smaller operators took over and stump farmers became proficient at bush work and the art of broadaxe use, and in a few cases were able to finance their own smaller operations instead of "sub-contracting".

6

Hauling

All tie hauling was by horses and sleighs during the 20's and well into the 30's, as horses were the only source of power available. As with the making of the ties, there were two types of hauling procedures—the stump farmer and the commercial type operations.

From the start, the stump farmer's first priority was a team of horses and the necessary rigging. If and when he was able to have his "druthers", his horses would weigh about 1200 lbs. each. Animals of this weight were versatile, light enough for every day driving and ordinary farm work and heavy enough for limited bush work, and what was most important, took less feed to winter than did the heavier types.

The tie hauler's (Stump Farmer or professional) tool requirements were few. A picaroon to use for pulling the ties on and off the tie rack, an axe for emergency use, and a Dryass...a heavy gunny sack kept filled with dry hay used for the important function of keeping the driver's... Oh, you guessed it... dry. The camp teamsters and most stump farmers had some type of blacksmith outfits and did their own horseshoeing. Sharp shoes on horses were as necessary for winter work then, as snow tire and chains are today for winter truck logging. In the rougher going, the ultra sharp shoes took a heavy toll of the horses in the way of leg and foot cuts, and

Late 1920's, near Smithers. Oscar Lundstrom at his tie camp displays tiemaking tools. Note building corners. The logs are "spiked" to board corners rather than notched together. This type of cabin could be erected very quickly. [Photo courtesy Mrs. Mabel Riley]

healing these deep cuts after they were sewn up by ordinary needle and catgut was often a problem. Many medications were experimented with, and what proved to be the most effective was a home type remedy. The fluid from balsam tree "blisters" was collected and mixed with lard. Simple, but cheap and always available.

The stump farmer was content to haul small loads, usually 30 to 40 ties and sometimes only one load per day, as his hauling was usually confined to his own cut of 1,000 or so ties per season. Occasionally he might haul for a neighbour who did not have a team or even haul for a commercial operation if his horses were good enough, but usually he was content to just handle his own cut, since this would usually mean that he could be home at nights with his wife, family and beloved stumps.

The hauling from the commercial camps was usually on a piece work basis, the operators providing horses, feed etc., and paying the teamster so much per tie usually 5¢. They used heavier horses (1300 - 1800 lbs.) and a top-load system. The loads hauled out of the cutting areas were "topped off" from piles previously hauled out to the start of the main road. This short distance hauling of the topping off ties was done with a light team by the "road monkey", a roustabout type who also travelled and maintained the main haul road.

About 5 miles was the hauling limit, as economics called for a minimum of two loads per day. When the teamsters encountered trouble their work days were long.

Although the contract qualifications specified "A maximum face of 10 inches", the ties produced in the

1920's, Smithers [Driftwood area]. Horse barn at Oscar Lundstrom's tie camp. Usually the very minimum of effort was put into the erection of the expendable tie camp buildings. Note "dryass" on sleigh. [Photo courtesy of Mrs. Mabel Riley].

1920, Topley. Lead load, Eric Strimbold. The first tie hauling was by the "lengthways" method. The narrower loads made for less road cutting but loading was difficult and "crossways" hauling soon took over. [Photo courtesy of Mr. S. T. Strimbold]

Early 20's, Smithers area. Building the "lengthways" load.

early 20's were mostly used for immediate replacement purposes, and the maximum size specifications were not enforced by the inspectors. This resulted in a large, heavy product averaging about 300 lbs. per unit. The average load from the large operations was 60 of these units—9 tons. As with the making, there were outstanding feats claimed. I saw 94 ties hauled into Rose Lake landing on one load, and have heard stories of much greater achievements—up to 200 in one load.

Some braking system was necessary to help control these huge loads coming down the hills. On the steepest grades a "roughlock" (a heavy chain) was wrapped around the sleigh runners. It could be easily freed when required by releasing its end from where it was held by a "fingerlink" fastened to the tie rack.

On the longer, gradual grades, old hay or manure was strewn at intervals to prevent the loads from getting too much momentum. An elaborate mechanical system was also used in a few cases. Sets of curved iron "dogs" which straddled the rear runners were controlled by a rope which ran through a pulley hung from a pole extension at the rear of the sleigh. The rope then returned to another pulley at the front of the sleigh, from where it was controlled by the driver. Theoretically, when he pulled on the rope, the "dogs" dug in and slowed down the load in proportion to the pull used. It did not work too well in practice, however (the driver didn't have enough hands to control his horses and operate the braking system), and its use was very limited.

The horses used by the large operation were turned out on the range in the spring, their work completed until the next season. The stump farm horses were not quite as fortunate, but they too, spent much of the off season free, since few of their owners had fenced pastures in which to hold them between farm jobs.

So during those early summers, herds of up to a hundred horses of varying colors and sizes were a common sight in the lush, practically untouched range lands of the Bulkley Valley and the Lakes District.

Cinched down and ready to roll. [Photo courtesy of Mrs. W. H. Morris]

1924, Decker Lake. The Saul brothers, Gin [left] and Bob [right] are taught the correct way to unload ties by local school-ma'am Margaret Palmer. [Photo courtesy Mrs. E. F. Mould]

1923. The winter's record load. [Photo courtesy of Mrs. E. G. Strimbold]

1923. Four-horse teams were not used in the actual bush areas. The extra team is used here to help "break out" new road. Note especially the brake lever arm at rear of sleigh. [These brake systems did not prove practical—see text]. [Photo courtesy of Mrs. E. G. Strimbold]

1928. Stump farmer Art Bach brings a small "jag" down the Rose Lake hill to the landing. [Photo courtesy of Mrs. E. F. Rowland]

Early 1920's, probably Quick, or Walcott. "Slooping" ties over bare ground for direct loading. This settler type operation was only possible when timber was very close to the landing. Even these relatively small loads could only be hauled a short distance over bare ground. [Photo courtesy Mrs. W. H. Morris]

7

The Loaders

Prior to about 1927, all ties went direct from the cars on which they had been loaded to the railway grade for replacement purposes, and maximum size specifications were not enforced by the inspectors. The average weight of these oversized ties was around 300 lbs. and loading them was a formidable task.

Unpeeled, untreated ties had a useful life of less than 10 years, so after the Railway had met its initial replacement needs, a creosoting plant was established at Edmonton, and most of the district's production after 1927 was shipped there for treatment. A creosoted tie lasts 2 to 3 times as long as an untreated one.

This innovation had a pronounced effect on the

industry. The ties now had to be peeled. This was a slow difficult job in winter time, and in severely cold weather often took longer than the actual hewing. This was an unpalatable pill to many professional tiemakers, and rather than swallow it they left the industry. In addition, ties with more than a 10" face were required to be "squared" to 9". This meant that most No. 1 ties were now 9" x 7" x 8' squared timbers. While this was not appreciated by the makers (and hastened the inevitable switch to sawmilling) it was a boon to the loaders, as it brought the average weight per unit down to about 200 lbs.

Even after this change, loading remained a demanding task that only the toughest and strongest of men could handle. It was a simple uncomplicated procedure. The loader upended the tie, bent low enough to have the longer (and heavier portion) above his shoulders, then grasped the thing firmly with both arms, straightened up, and walked to and over a 16" plank into the boxcar, the tie balanced precariously upon his shoulder.

They were paid 3½¢ to 4¢ per tie and usually loaded a car (400 per man each day). The wages of $14.00 to $16.00 per day made them by far the highest paid workers in an era when going wages were 35 to 40 cents per hour, and a team of horses and their driver could be hired for $8.00 a day. Required equipment was very simple—a shoulder pad and a picaroon, so why the fabulous wages? A matter of supply and demand, for even in an era that produced strong hardy men, very few could meet the physical demands of tie loading.

Consider the statistics. The average carry from pile to car was about a 100 feet, and the average tie

Early 1920's, Topley. Stan Horning carrying a big one out to the haul road. It was estimated to have weighed over 350 lbs. [Photo courtesy of Mr. S. T. Strimbold]

weight 200-300 lbs. So, during a day's work of loading 400 ties, our loader travelled about 15 miles, half of this distance with a 200-300 lb. tie on his shoulder. His daily lift and carry was about 20 tons.

A crew consisted of four loaders and an inspector. They travelled from siding to siding by whatever means available. Sometimes a stump farmer would drive them to the next siding, sometimes it was a section crew hand-car and in a pinch they toted their tools and blankets and walked.

The inspector, a railway employee, kept a record of grades and timber marks, and the contractors received payment on the basis of his reports. He was referred to as "God's Brother" by the stump farmer

1923, Decker Lake Landing. The centre loader is Otto Lundberg who loaded ties for over 20 years and became a legend in his own time. The big man is Tie Inspector "Tiny" McLaughlin, also a very well known character during the hewn-tie era. [Photo courtesy of Mrs. E. F. Mould]

Circa 1922, Topley. Ovie Watson, Ralph Smith and Laurence Kylling "do their thing". [Photo courtesy of Mr. Ivan Holmes]

subcontractors, some of whom, not being expert broadaxemen, could be badly hurt financially by a tough inspection. Actually, the inspectors had little leeway after the ties started going to the creosoting plant. They had to meet the rigid specifications or be "culled" or the inspector would be out of a job. In some cases, a soft hearted inspector would delay the loading of a poorly made stump farmer's cut until an order for ties for immediate replacement purposes was received.

The loading crew had to find board and sleeping accomodations as they moved along the line, but this was seldom a problem. Because of their high earning power, it was considered in order to charge them 50 cents per meal (the going hotel rate), and a dollar per night each for sleeping accomodation, and "boarding the loaders" was a bonanza for the stump farmer's wife who could land it. The $3.00 - $4.00 a night for sleeping room was all clear profit, and a "hell" of a lot of money in the early 20's.

Because of the nature of their work, the loaders' appetites were enormous. On one occasion, the first morning after their arrival, a young homesteader's wife decided to start their breakfast with porridge. She set out the filled family bowl and turned to the cupboard for the individual dishes. Imagine her consternation on returning with them to find that the first man in had liberally milked and sugared the main bowl and was enthusiastically gulping the contents down.

The tie loaders were the elite of an industry that was composed of strong, skilled hard working men. They vanished from the scene many years ago, and it is highly improbable that their likes will ever be seen again.

8

Fluming

During the summer of 1918, a disastrous forest fire burned over a vast area about 25 miles East of Burns Lake. In one portion of the burn a large volume of very good tie timber was scorched, but remained salvageable if harvested immediately, before "checking" could occur. The timber was made available for immediate cutting at a minimal stumpage rate and the Hanson Lumber Co., ever alert to the possibilities of a successful operation, sent out men to evaluate the situation. They decided the seven or eight miles to the railway was too far for a profitable sleigh haul operation but recommended an alternative.

The timber was in the vicinity of Shovel Lake which was drained by a creek of the same name down

1919. The railway spur near Priestly showing the flume which transportated ties from the creek to the loading area. Here they were pulled from the flume and loaded into the box car. [Photo courtesy of Smithers Museum]

to the Endako River, near the railway station of Priestly. The foresters felt that the ties could be economically waterborne to the loading spur.

The Hanson Co. acted on this report immediately and by fall a crew was in the area to cut a "tote" road, build camps, clean out the creek, and construct about a quarter mile of flume required to lift the ties from the creek to a hastily built railway spur.

In the early fall, a large crew of tie hackers was engaged, (they received a bonus for working in the blackened timber) and the winter of 1918-19 saw more than 150,000 ties hewn and skidded to the lake and creek banks. In the spring they were floated down Shovel Creek, diverted into the flume from

1919. Hewn ties piled along Shovel Creek ready to be watered for the seven mile trip to the flume at Priestly Spur. [Photo Courtesy Smithers Museum]

where they were loaded directly onto railway flat-cars. The Railway required the ties for immediate replacement use as road bed maintenance had been badly neglected during the immediately previous war years, and they co-operated by keeping a switching unit on hand throughout the driving and loading period. For about a month in the spring of 1918, a train of 25 flat cars, each loaded with about 350 hewn ties, left Priestly Spur every day. The Priestly operation was among the first, and was by far the largest, of the waterborne type operations. More than half of the 270,000 hewn ties produced during the 1918-19 season came down Shovel Creek and through the flume.

A few years later, when suitable timber stands of large enough volumes to ensure profitable sleigh-

1919. DISASTER. The driving and loading operation was synchronised as well as possible, but a breakdown did sometimes occur. With no communication between those watering the ties and the loading crew, and no holding space in the small river, when the loading "foul up" occurred, these ties were carried on into the timber at the end of the flume. [Photo courtesy of Smithers Museum].

haul, commercial operations became scarce, the Hanson Lumber Co., remembering their success at Priestly, began looking around for similar possibilities.

Their cruiser located a good stand of timber in the area of the confluence of Dungate Creek and the Buck River about 5 miles south of Houston. Here, again, they decided that skidding to the creek banks, driving the river, then fluming out to the loading area would be more profitable than expensive road making, and the long sleigh haul. There were some obstacles not encountered at Priestly. First, the rapid spring run-off resulted in too short a period with sufficient water for a successful co-ordinated driving and loading out operation. This was solved by damming a large meadow area at the headwaters of the Buck River to hold and control the spring run-off. Second, there were some short canyon sections which were too fast and rocky for the ties to pass through without being damaged and it was necessary to build flumes or chutes over them, the ties safely speeding through these areas some distance over the river bed.

The final problem was to get the ties out of the river bed to the loading spur. The carrying flume had to pass OVER the existing highway, so it was necessary for the flume to start a considerable distance from the loading spur for the water to retain enough velocity to take the ties to their final destination. The intent was to synchronise the driving and loading as was successfully done at the Priestly operation, but there were too many variables and not enough volumes to warrant the Railway keeping a switcher engine at this operation. As a result, there were some hectic experiences when the loading facilities bogged down and the ties kept right on coming and coming. There were no communication facilities between the "dumpers in" at one end and the "takers out" at the other end, and at times it was necessary for a messenger to ride hell for leather up to the dumpers in and tell them "to cease and desist for a while". In spite of the obstacles, over 40,000 ties were successfully brought down the river, through the flume, over the road and loaded out in the Spring of 1930 but this was the last of the fluming operations, although lakes and rivers were to play a large role in the transportation of ties.

9

Lakes and Jackladders

At first all the ties were hauled into the railway sidings for loading on to the rail cars, but piling space at the sidings was limited, and when large scale operations spread to the south side of Burns and Decker Lakes, and in 1922, extended into the Francois Lake area, a new phase of the industry came into being—water transportation and jackladder loading. The jackladder was a structure supporting a power driven endless chain or belt, fitted with holding lugs at appropriate intervals, which lifted the ties fed on to it from the lake, up to the box car on the siding. A short hinged section extended about a foot into the car. Here, four men (two to a tie), alternated in grasping the dripping ties as they came off the belt, and stacked them in the car.

The crew was composed of the four men in the car, a couple of boom men feeding the ties on to the jackladder, the inspector—and a horse. As it was not feasible to move the jackladder, each car had to be moved as loaded, the first of the string far enough to allow the rest to be moved along as they, in turn, were loaded. This was the job allocated to the horse—usually a heavy strong animal, pensioned off from bush work.

The men working in the cars received 1½ cents per tie which they split four ways. They loaded about ten cars per day (3-4000 ties) thus averaging about $10.00 per day each. This was big wages for the times, but far less then the backpackers received. The cost per tie of jackladder loading was considerably less than the back packing method, but it was only at Decker Lake, Rose Lake and Fraser Lake that the combination of available timber and proximity of lake to rail siding necessary for jackladder loading was available.

The ties cut at Francois Lake were hauled to the lake shore during winter, then after breakup, boomed, (in the case of small sub-contractors, the boom was towed into their landings, opened up and the ties checked into it), and towed to the outlet of the lake. There they were channelled into the Stellaco River, floated 10 miles downstream to Fraser Lake, reboomed, and towed to the rail siding on the north shore where the jackladder was located. The annual load out involved many thousands of ties, as it included all the Francois Lake cut, and a large volume from the immediate Fraser Lake area.

The south side of ten mile long Decker Lake was covered with high quality tie timber which attracted many operators during the early 1920's. At first, the ties were hauled across the ice, but pulling the huge loads from the lake up to the siding was a difficult problem and piling space was limited. The hauling distance, too, became a problem as the close-in timber became depleted. So it was inevitable that advantage would be taken of rail siding and lake proximity to install a jackladder. The lift involved presented a challenge but the versatile tie men met this successfully and tie booming and jackladder loading became an important seasonal part of the Decker Lake economy for almost two decades.

A few miles to the west at the tiny settlement of Rose Lake, over 30 teams were hauling into the landing during the first years of tie operations, and it was a shortage of piling space which prompted a short lived but disastrous jackladder loading operation. It was the year (about 1925) that the contracts first called for the ties to be peeled and the contractors, who were piling the ties on the little lake for later jackladder loading, decided to leave the peeling of the ties until after breakup. Then, rafts were built, the now floating ties pulled on to them, peeled and then slid over the other side. Then they were jackladdered out and it appeared the operation was highly successful. The contractors had avoided the problem of getting the ties peeled during the cold winter weather, hauling was easier with the bark on, local peelers paid by the tie made big wages, and loading costs were lower. Why then was the operation a disaster? Well, the bark from 60 to 70,000 ties which was left in this beautiful lake polluted the water and totally destroyed the fish population, which had included rainbow trout up to five pounds.

Early 1920's, Decker Lake. The winter's "cut" heading up the jackladder to the waiting box cars.

Early 1920's, Decker Lake. Many thousands of ties were hauled across the ice of Decker Lake. [Photo courtesy of Mrs. C. Paulson]

So, there really is nothing new under the sun—it was over 50 years ago that the first payment was made to "the price of progress" in this area.

The south side of Bulkley Lake also produced large volumes of ties during the early years, but although the railway runs adjacent to the north side for the four mile length of the lake, conditions were not favourable for a jackladder operation. Most of the ties were hauled over the ice and then up to the Rose Lake siding. Others were watered, towed to the foot of the lake and floated about a mile down the Bulkley River to "Hultkrans Spur" where they were taken out and loaded on to cars by back packers.

The area adjacent to Tchesinkut Lake was also rich in tie timber, but the ten miles to the railway was too far for sleigh hauling and it was not until 1928, when truck hauling was introduced, that the first tie cutting occurred in this area. The ties cut along the lake were boomed and towed to where the Burns-Francois Lake road skirts the west end of the lake. Here, a modified type of jackladder was used to get them from water to truck for the 10 mile haul to Burns Lake.

Hewn tie operations were never attempted in the Babine Lake area, but after the arrival of portable sawmills in the 1930's, ties were sawn on the east side of the lake, watered, and towed to Pendleton Bay, jackladdered on to trucks and hauled 22 miles to Burns Lake. The double handling and long hauling distance involved proved too much for financial success, however, and this operation was very short lived.

The use of lakes as winter haul roads, springtime water transportation and jackladder loading were all colorful phases of the hewn tie industry, that were not without their dramatic incidents such as an event which occurred in the early 1920s, when a sleigh loaded with people, including a young mother and her four small children were nearly drowned in the icy water of Decker Lake.

They had driven across the lake from a tie camp, on an extremely cold December night to attend the annual Xmas Tree concert and dance at the Decker Lake schoolhouse.

Soon after midnight the now sleepy youngsters were bundled into their heavy winter clothing and loaded into the sleigh, happily clutching their Xmas presents. The adults said "Merry Xmas" to their friends and the team, anxious to reach the warm, friendly confines of its home barn, started with a dash on the return trip over the Lake which is about two miles wide at this crossing-point.

About fifteen minutes later, those still at the party (now a dance) heard yells for help echoing across the cold clear night air.

The men immediately realized what had happened, and paused only long enough to pull on their outdoor winter footwear and grab gas lanterns and ropes before dashing down to the lakeshore and across the iceroad towards the sound of the cries, which suddenly became ominously silent.

Arriving at a gaping area of open black water, their worst fears were realized. It was evident that the outfit had broken through the ice, sleigh and horses were gone and there was no sign of life,—nothing but the mute, unmistakable evidence of miscellaneous debris floating in the already refreezing opening. For a few moments they stood in stunned silence, gazing at the sombre scene.

Then someone saw it—a light which suddenly appeared in the area of the camp buildings on the lakeshore about a half mile away.

"Let's go. Somebody has made the shore." one of the party shouted. Arriving at the camp area within a few minutes, they were overjoyed to find the entire party safe including the youngsters, who had barely awakened during the whole affair and were already asleep in their beds.

The horses had dropped through the ice without warning but the sleigh had remained up just long enough for the children to be tossed to safe ice and the adults to scramble after them. The mother and some of the others had then immediately grabbed up the youngsters and headed for the camp. Some remained to make a valiant effort to get the horses out (this was when the calls for help were made), but within minutes the icy water took its toll and the animals and sleigh disappeared into the depths. The men then became aware of their own danger and wet and freezing clothing and they too made a dash for the shore and warmth of the log cabins that were the tie camp.

Actually it was not an entirely unusual event for an outfit to be lost through the ice during this era of extensive winter lake hauling, for although by mid-winter it was thick enough to support huge loads, the weight of accumulated snowfalls could bend it down until the water below was forced up through the ice, the weakest spot giving way to the inexorable pressure.

This phenomenon results in the development of what is termed an "air hole". The area around the depressed area is flooded and the ice in the immediate area weakened so that, although the average air hole is only a foot or two in diameter, a much larger area is unsafe until it refreezes.

This is what happened on the night of the near tragedy. A check the next day revealed that an air hole had developed overnight within a foot of where the party had travelled safely on their earlier evening trip over the lake.

What set this break through the ice apart from other such disasters of the era was the presence of the small children, and the fact that the trip was made at night. People did not normally travel over frozen lakes after dark. To do so in view of the ever present "air hole" danger was considered crowding your luck, but this particular night was the occasion of the community Xmas tree party, which in those long ago days everyone within horse driving distance attended.

The era of lake travel had many dramatic incidents but it is doubtful if there was ever a closer escape from death than was experienced by those pioneer tie makers, and a mother and her four babies on the ice of Decker Lake in the early 1920's.

10

Travel and Transportation

The early settlers had travelled hundreds, and in many cases thousands of miles with a common goal in mind—to obtain land, build homes, and establish a new way of life. Further travel, except as it related to these initial plans, held very little interest for them.

So, at the start, the fact that they were virtually isolated from the civilized world, was of little concern. There was the railway link, of course, but the three passenger trains per week, seldom operated on schedule, and apart from bringing in mail, played a negligible part in the affairs of the busy homesteaders. Their association with the railway was generally limited to the local "way-freight", which provided a weekly service between all "way" points,—the railway stations situated at

Circa 1920, Smithers area. Speed and power a-plenty here. [Photo courtesy of Mrs. W.H. Morris]

intervals of six to eight miles—to pick up and deliver commodities of every kind. This service was the life line of the scattered population.

The way-freight left Smithers in the early morning and when it reached Endako (or Pacific on the westward run) was governed, not by a preset schedule, but upon the amount of business transacted enroute. This was mostly pickup and/or delivery of food supplies, farm produce and equipment, hay and oats for horses in the tie camps—in fact anything and everything involved in developing a new frontier. During the tie loading season the way-freight engine also did the switching at the sidings, taking out the loads and spotting empties,

Circa 1920, Decker Lake. Ready for a quick trip to town. [Photo courtesy of Mrs. C. Paulson]

and throughout these periods, many long hours were put in "waiting for the way-freight" by people expecting perishable goods.

The railway stations were the focal points of the little communities and the surrounding areas and the settlers were linked to them and thus with each other by roads they cut out themselves. Travel over these narrow stump-filled trails was by horses and wagons and was kept to a minimum.

The right-of-way of the ill fated Collins Telegraph Line, abandoned immediately after it was put through in 1866, then reactivated at the turn of the century, had been used intermittently over the ensuing years as a travel route by natives, gold

1922, Walcott. All dressed up and waiting for the train. The watertank supplied water to the steam locomotives of the era. The tanks were about 25 miles apart and water was pumped into them [see pumphouse at right]. It was fed into the locomotives water tanks by gravity. [Photo courtesy of Mrs. W. H. Morris]

seekers, packers, the first pioneers and, when practical, as a tote road during the railway construction. In the early 20's, although still little more than a trail, it provided the base of the rough road which linked the little settlements between Hazelton and Burns Lake.

There were a few cars within the area (mostly at Smithers, Burns Lake and Hazelton) prior to 1919, but as there was no road west of Hazelton and no access to the existing roads east of Burns Lake, their use was very restricted. Within a few years,

1926, Telkwa. "Model T's" were taking over, but one almost expects to see Matt Dillon come a-riding round the bend. [Photo courtesy of Mrs. W. H. Morris]

however, the settlers were enjoying what was to be a brief period of prosperity. The better class land was starting to produce, and the stump farmers (those who had "toughed it out") had mastered the use of the broadaxe and were making a comfortable living from the tie bush and their homesteads. The link from Burns Lake to the existing road system in the east was completed in 1924 and soon after this event automobiles made their appearance in numbers throughout the entire district.

This was the heyday of the Model "T" Ford, and within a surprisingly short time it seemed that every settler in the district owned one. They were marvels of simplicity in construction and operation, and in a do-it-yourself era their owners found maintenance and repairs a relatively simple matter. Few of these cars ever saw the inside of a commercial garage.

They were actually driven very little by today's standards, mostly for short trips within and between the little communities. A trip from Burns Lake to Smithers took a long hazardous day and such trips were usually made only for special or emergency reasons. The writer and some friends made a trip by Model "T" from Burns Lake to the harvest fields of southern Alberta during 1927. It took ten full days (longer than it now takes to reach the moon) and was an adventure still remembered with awe and wonder.

During the winters, the roads were kept open by horses and sleighs, and even the versatile Model "T"s could not negotiate the resultant narrow trails. In any case, there were no car heaters, no anti-freeze and of course, no "plug-ins", so in the fall most owners "stabled" their cars and turned back to horses for winter transportation.

By the start of the 30's, car use had increased considerably as the road was improved and snow ploughing programmes were initiated, but the first automotive era was short lived. The Depression arrived and car operation was among the first things affected. It is probably difficult to appreciate at this date but in an amazingly short time few people had money with which to buy gas. For a while, the cars just seemed to sit waiting for the Depression to go away, but when it refused to do so, the owners just shrugged their shoulders—they had lived without cars before and could do so again—and turned back to their horses for transportation.

Some cars survived the Depression, others just seemed to disintegrate over the years. Many were stripped down, wagon poles fitted and became the infamous "Bennett buggies". Travel outside of immediate local areas was again pretty well restricted until the railway people came up with an innovation. They put a passenger coach on the tail end of the way-freight, which was still making its weekly trip up and down the line.

Travelling by way-freight coach was a nerve wracking experience; there was no service, no heat, and of course, no schedule. The only time passengers saw a trainman was when he collected fares and often even this function was overlooked. During tie loading periods, a traveller might wait at, say, Houston for several hours to board the train for Burns Lake, then ride as far as Topley (20 miles), wait there a couple of hours while the engine "switched" tie cars and repeat the experience several times before reaching Burns Lake six or seven hours later.

It was during this period that Burns Lake hosted an important ski meet, and as they had an outdoor rink and a hockey team of sorts, decided to stage a hockey game as an added attraction. I was one of a group of young men living at Rose Lake, who had a rink on the lake where we skated and played at hockey among ourselves. To our surprise, we were invited to be the opposition in the big hockey game, which was scheduled for 2:00 p.m. on a way-freight day, and a very efficient Tournament committee informed us that they had prevailed on the C.N. people at Smithers to start the train from that point

Early 1920's, Telkwa. This was the main highway immediately east of Telkwa. It hugged the bank of the Bulkley River for several miles eastward. [Photo courtesy of Mrs. W. H. Morris]

1927, Rose Lake. Even the "Model T"'s required chains to travel the highway at times. [Photo courtesy of Mrs. E. F. Rowland]

early enough to pick up the Rose Lake team and get them to Burns Lake to meet the starting time deadline.

An excellent arrangement, but there was a slight problem. This was the Dirty Thirties and there can be no better illustration of how dirty the thirties really were than the stark fact that we had no money with which to pay our fare, a matter of about 30 cents each. Not daunted, we were on hand, hidden from the crew's sight, when the train lumbered to a stop at the water tank and quickly threw our scanty gear into an empty box car and scrambled in ourselves without being spotted. The train moved up so that the

1923. A trapper ready to head out on his initial trapline trip. The packhorse will only be used on the first trip to hold take in traps etc. [Photo courtesy Mrs. E. F. Rowland]

passenger car was at the platform and we watched through a crack in our box-car door as the trainmen paced up and down waiting for the hockey team to show up. After a long wait, we heard the conductor tell his crew "Well, we got here in plenty of time—we've waited a long while and still no sign of any hockey team. To hell with them—Let's go." and away we all went and got to Burns Lake in time for the game to start on schedule. No, we didn't win. This was our first (and last) game and we were bombed. Still, it wasn't entirely a lost cause. We probably still hold the record of being the only hockey team in history that "beat our way" to a game on a train that was especially scheduled for us.

It was not until after the start of war in 1939 that economic conditions improved and automotive travel again took over, this time to stay.

11

Rabbits

During the initial years the stump farmers lived off the land to a considerable degree. Wild berries were plentiful, and were picked and preserved to augment food supplies. Huckleberries were the most sought after. They grew in profusion at many of the higher levels, there being a local "huckleberry mountain" within reach of most little communities and often family groups would travel by wagon or packhorse on berry picking expeditions. The natives, too, harvested berries during the picking season as they had been doing for countless generations but there was no conflict. The settlers were few and there was plenty for all. For years the natives sold four gallon gasoline cans full of huckleberries for 50¢ a can. During the Depression years, they found few buyers even at that price, but soon after the outbreak of War Two our present inflationary period started when, without a blush, an elderly Indian demanded $2.00 for his can of berries. When the startled buyer asked why the sudden raise in price the old native pointed eastward and solemnly said, "Big war over there some place"—and that was that.

Fish too were plentiful, trout filled lakes being within reasonable reach of most settlers. The lakes are still there, of course, but alas, most of them have been "fished out" now for many years.

Big game, with the exception of a few deer, was practically non-existent in the early years. There were some caribou and goats at the higher levels but these were far out of the reach of the average stump farmer. There were no moose until they slowly started to drift into the area during the mid 1920's. The first moose I know of being killed in the upper Bulkley Valley summit area was shot near Palling in 1924. The first I saw were two animals near Perow in 1925. Soon after this, moose sightings became a common occurrence, and by the start of the 1930's the animals became firmly established as a major part of the area's wildlife population.

Grouse and wild waterfowl were plentiful during the early years and also formed an important part of the struggling settlers' meat supplies. We raised some tame ducks at Rose Lake, which would take off up the lake and not return until dusk. One Sunday, the local section foreman dropped by and gave my mother four ducks he had shot. A neighbour showed up a little later and, seeing the ducks, scoffed "They're fish ducks and not edible", so they were relegated to the garbage pile. At dusk, the tame

1923. This sad looking boy lost his mother about the same time the baby owl [on fence] lost its parents to bounty hunters. So they adopted each other. [Photo courtesy of Mrs. E. F. Rowland]

1925. Early Topley settlers all set to start on a berry-picking safari to Topley's "Huckleberry Mountain".

ducks straggled home, minus four of their number. A check revealed the horrible truth. The "fish ducks" were those missing from our tame flock.

The animals which were truly the mainstay of the early settler's diet and contributed more to his meat supplies than all other sources combined were the lowly rabbits.

Rabbits are mentioned by most of our early historians as being one of the main sources of the native's food for as far back in time as such knowledge is available. The word "Kitwanga", given to the Indian village on the Skeena River means "place or people of the rabbits". Until the white man's interference with nature's delicate balance, the rabbits had recurring cycles of population increases, which "peaked" at about seven year intervals. During these peak periods, their numbers reached epidemic proportions but, just when it seemed they would inundate the entire universe, disease would infect them and millions would die off. Then, the disease would run its course, leaving only the stronger animals, and the cycle would resume. Rabbits breed several times a season, bearing two to four to each litter and, as they seldom travel far from their birthplace, it seems reasonable to assume that as their numbers increased, inbreeding weakened their resistance to disease, nature thus providing its own control mechanism.

The rabbit population was at the peak of its cycle during 1921 and 1922 and this proved to be a blessing to the stump farmers, who snared the animals by the hundreds for table use and, in some cases, as food for pigs.

Rabbits are creatures of simple fixed habits. They

1920. Children also had miniature trap lines. These lads are about to traverse their 3-mile line in the Rose Lake area. [Photo courtesy of Mrs. E. F. Rowland]

1920. The rabbits were the basic food supply for large numbers of fur-bearing predators. Pictured is the winter's fur catch of an unknown trapper. Smither's area. [Photo courtesy Mrs. W. H. Morris]

locate a hideout, which they leave only at dusk and daybreak to travel to the nearest food supply (grasses in summer, the bark of willows and small trees during winter), where they feed in the short periods of half light between sundown and darkness, and dawn to sun-up, returning to their lairs to patiently sit out the daylight hours. They always use the same routes and their trails become narrow, hard packed paths, parts of which might be used regularly by many animals. So, snaring the rabbits was a very simple procedure. Merely, to hang the right sized noose, fashioned from a fine wire, over the trail. Some would bend an adjacent sapling over, lightly fasten it down, and attach the snare. When the rabbit entered the snare, its' death throes would release the sapling which would spring back erect and the victim was more or less "hung up to dry".

There were millions of rabbits during the peak years. Many of the new homesteads were closely hemmed in by the thick bush in which the animals flourished and when the farm stock was fed at dusk, hundreds of shadowy, ghostly little white forms would hippity-hop across the little clearings to join the other animals at the hay piles.

One winter, during a "silver thaw" period, (a winter rain, followed immediately by sub-zero weather which encases the entire forest growth with ice), it was impossible for the rabbits to get food and millions perished of starvation. The hardier ones broke their nocturnal feeding pattern and lost all fear of man in their endless quest for food. When tie makers felled their trees, the brittle tops would shatter, shedding the ice. Even in broad daylight, rabbits would appear, as if by magic, and it was a common occurrence for a man to be hewing the butt of a tree while dozens of rabbits fed fearlessly upon the shattered top.

As the rabbits flourished, they provided food for hosts of predators, mainly coyotes, lynx and weasels whose hides in turn contributed in varying extents to the livelihood of the stump farmers, many of whom operated short traplines. Often the children, too, would run miniature lines, and many of the simple Xmas presents exchanged during those early years were paid for with money received for weasel skins.

Among those who preyed upon the rabbits were the owls—thousands of huge horned owls. During the early 1920's, if a traveller stopped anywhere on the trail from Hazelton to Burns Lake, he would be sure to hear two sounds. Throughout the day the rythmic k-chuck, k-chuck, k-chuck of tiemakers' axes, and at dusk, the hoot-ta-hoot-hoot and answering Whoo-whoo of the great horned owls.

During 1923 or 24, an amazing event occurred. The wildlife administration of the day, which had for some time payed a bounty on coyotes and wolves, reached a new low in ineptitude when they placed a Government bounty of $1.00 per bird on OWLS.

Any opportunity to augment their scanty cash income was seized upon avidly by the stump farmers and the owls immediately were shot and trapped by the thousands. How does one trap an owl? It's amazingly simple. You nail a board, just large enough to hold the trap, onto the end of a pole, find a clearing in a rabbit feeding area, and raise your pole with the set trap upon it. The owl is always searching for such a perch in such a clearing. He spots the pole, alights on it and the trapper has earned another dollar. One family in the Francois Lake area amassed enough of these dollars to purchase their first Model "T" car.

The stupid and senseless near annihilation of the horned owl population seemed to upset nature's delicate balance and probably played a major role in the breaking of the age old pattern of recurring cycles of large rabbit populations.

In any case, that phenomenon, along with the stump farmers and broadaxemen, appears to have passed from the scene forever.

12

The Log Schoolhouses

Next to making a living, the most pressing problem the post War One settlers faced was to provide basic education facilities for their children. They quickly learned that the B.C. Government of the day did not concern itself to any great extent with the building and administration of remote rural schools, per se. The Department of Education told the anxious parents, in effect,

"If and when you can guarantee a starting attendance of ten pupils, a school district will be designated to cover your local area. The property (land) held within its boundaries will be assessed taxes which we will collect and use to pay your teacher's salary, for as long as the average school term attendance is not less than six pupils, and the

1920, Decker Lake. Fred Smith, Charlie Saul, Henry Piche and Bill Featherly start construction of Decker Lake's first school.

A few weeks later: The builders and their families proudly pose in front of the completed building. [Photos courtesy of Mr. L. G. Saul]

registration for the ensuing term is a minimum of eight. All other problems relative to establishing and operating a school, including providing and maintaining the building(s) will be your headache, and you must elect a three man school board committee from among your number to co-relate your efforts. Good luck to you. You are going to need it."

The settlers met this challenge head on, in some cases meeting the initial problem of providing ten starters by temporarily "borrowing" children from prolific families in neighbouring communities. Between 1919 and 1924, many miniature school districts were established and primitive one-roomed log structures erected to serve them as schoolhouses. Nine of these covered the area between (not including) Telkwa and Burns Lake, a distance of about 80 miles, and their registration figures for the 1923-24 term are still on record as follows:

	Children registered	Families represented
Woodmere	8	4
Round Lake	9	5
Quick	10	4
Houston	14	7
North Bulkley (Perow)	11	4
Topley	16	7
Rose Lake (Omineca)	12	6
Palling	10	4
Decker Lake	11	5
Total	101	46

Authentic figures for other areas of the Bulkley Valley-Lakes District are not available to this writer, but it may be safely assumed they were generally comparable and a reasonable estimate of the total school attendance for the area now covered by the Burns Lake and Smithers School Districts would be less than 250 pupils in 1924. Fifty years later, (1974) the official registration from the same area was 4803 pupils.

The school buildings were all erected by volunteer efforts. At Rose Lake, where I was raised, six sets of parents, their older children, and a few bachelors cut and skidded out the logs with which they built their schoolhouse walls, and also took out timber from which they sawed out the lumber for the floor and roof with a home made sawmill, (the first used in the upper Bulkley Valley). The flooring they laboriously planed by hand. A loan of $300.00 was floated to pay for roof shingles, a door, and windows and within an amazingly short period of time another "little log schoolhouse" was ready for occupancy. With the possible exception of minor variations in detail, this was the pattern of how most of the first schools came in to being.

Surprisingly, it was not difficult to staff these isolated little schools, although at the start it was necessary to recruit the teachers from the far away, already developed areas of B.C.'s lower mainland and Vancouver Island. They were mostly girls of 18 to 21 years of age, anxious to get started in a teaching career. The profession, one of very few available to women at the time, was so overcrowded that it was a common occurrence for a rural school board to receive from fifty to one hundred applications for a

1923. Schools were the centre of community activity. An amateur "drama" group pose in front of their "theatre" the Rose Lake school. [Photo courtesy of Mrs. E. F. Rowland]

1923. Palling school children. Shown is the total enrolment and their number is typical of average attendance in many such schools throughout central B.C. during this era. [Photo courtesy Mrs. E. Schoof].

Palling School. A masterpiece of simple rustic art and also typical of log schoolhouses throughout the area. [Photo courtesy of Mrs. E. Schoof]

1928, Smithers [Driftwood]. Some teacher's boarding places were far enough from the schools to require transportation. Here a school ma'am and a pupil [Florence Lundstrom] start their daily journey. [Photo courtesy of Mrs. Mabel Riley]

position that would pay $90.00 to $100.00 per month and called for travelling 600 - 700 miles from civilization to teach 8 to 12 youngsters from Grades 1 to 8, in a one roomed log building which contained no plumbing or electricity; in fact, was heated by a wood burning heater which the young lady often had to light each morning, after walking from one to three miles from her boarding place.

There were some compensations, of course. Despite the primitive working and living conditions, the girls enjoyed a type of freedom that was not yet available to their counterparts in the more convention bound developed areas. They were often the only unmarried girls in the community, and much sought after by the unattached males. The fact that they had (in most cases) immediate ties with the far away cities clothed them with an aura of mystery and sophistication that fascinated and enslaved their rustic pupils to an extent that present day members of a mobile generation would find difficult to understand.

The girls were pioneers in their own right, and made an incalculable contribution to the development of the wilderness area. Few ever really looked back to what they had left, and many married fledgling homesteaders, remained to make a new home on the frontier and bear their own generation of pupils for the little log schoolhouses.

Grade eight was the highest taught in the rural schools and for the first two decades very few children received the opportunity to obtain any higher academic education. However, when they left the little schools, they were probably as well qualified to meet the challenges facing them as are the high school graduates of today's highly sophisticated school system.

The little buildings served the communities in many ways. Most of them doubled as churches, and in addition to regular Sunday services, many of the area's early weddings (and later baptisms) took place in them. They were used as meeting places, dance halls and general entertainment centres, even occasionally as a morgue when the deceased was considerate enough to "call it a day" at the end of the week or during a school holiday period.

The log schoolhouses flourished during the 1920's and 1930's. Most of them were replaced during the 1940's by frame structures, but these only served for less than a decade in most cases, until the upgrading of the area's road systems, and rapid improvement in transportation facilities brought about the introduction of the modern school bus concept, which, for better or worse spelled "finis" to the era of the one roomed schools.

1930's. A typical schoolma'am. Muriel Chapman came from Victoria to her first school at Rose Lake as a young woman.

1923 Rose Lake. Some kids walked a long way to school, and when the tie hauling started, the landings could sometimes be a good place to hitch a ride. Ted Rowland waits for neighbour Len Bedore to unload and will then ride the four miles home with him. [Photo courtesy of Mrs. E. F. Rowland]

13

Recreation

The terms "recreation" and "rest" were almost synonymous to the settlers during the initial years. In an era of extremely hard physical endeavor, their entire life style involved creating and "recreating". As they labored at the challenging task of building homes, schools, and making a living from the raw materials available to them; they had little interest in "recreation" as the term is generally interpreted today.

Physical "recreation", however, was a necessity, which most met by making Sunday a day of rest. Those with neighbours within reasonable walking or horse driving distance would interchange visits, and sometimes hold services of worship within their homes. This was about the extent of socializing and

Early 1920's. Everybody, including the Bach sisters, caught fish and the tackle required was very basic. [Photo courtesy of Mrs. E. F. Rowland]

The Telkwa Barbecue. The date this picture was taken is unknown, but it was probably previous to 1920. This is the oldest event in the Bulkley Valley and still gets bigger and better as the years roll by. [Photo courtesy Mrs. W. H. Morris]

Kids "skinny-dip" among the ties waiting to go up the jackladder at Decker Lake. [Photo courtesy of Mrs. C. Paulson]

Early 1920's. Meeting the trains was a social event. Here some sophisticated Smitherites await the arrival of the tri-weekly passenger train. [Photo courtesy of Mrs. W. H. Morris]

Early 1930's. "We made a rink on Rose Lake."

"recreation", until the building of school houses provided common gathering places.

As a midsummer diversion and an occasion to meet and gossip with more distant neighbours, most communities held an annual local picnic (the Houston May 24th celebrations go back to the stump farm days) where pickup and occasional intercommunity ball games were part of the day's Program. The Telkwa Barbecue, and Bulkley Valley Fall Fair, still going strong, were early annual attractions, but travel limitations generally restricted attendance to local area inhabitants during the stump farm-broadaxe era.

There was no organized athletic endeavor, with the exception of the Ski Clubs of Smithers and Burns Lake, who were intermittently active from the early 20's onward. This activity reached a high peak in 1932 when a member from the Burns Lake (Omineca) Club, Kaare Engstad, competed in the Winter Olympics held at Lake Placid, New York, and made a very creditable showing. Apart from this very remarkable achievement, competitive sports played a very small part in the lives of the stump farmers and broadaxemen.

The main source of recreation available to them can be summarized in four words—THE SATURDAY NIGHT DANCES. They were held in the schoolhouses on a rotating basis by mutual agreement among adjacent communities, so that there was either a local Saturday night dance, or one that could be reached by travelling from five to ten miles, almost every weekend.

It seems that there are, and always have been, musicians wherever people gather, for some kind of talent was usually available and freely donated by those who played for the dances, which started early and often lasted until dawn. Families attended en masse, (this was before the event of the "baby sitter"). The babies were put to bed among piles of blankets and coats, and usually slept soundly and peaceably through the din. The youngsters, from about four to twelve years, would participate for awhile, struggling to keep awake until the announcement for the "supper waltz" would indicate that their vigil was over. They would then stuff themselves with a amazing variety of cakes and sandwiches, brought by their parents, and in a matter of minutes join their baby brothers and sisters in sound sleep, bedding down wherever they could find space. They, too, seldom awoke, even when being bundled into sleigh or wagon box for the trip home.

The over twelve set would keep going. Youngsters learned to dance at a very early age and girls of twelve and over were in great demand as partners. The local schoolma'am, often the only unmarried adult female in attendance, was invariably the "belle of the ball", and from the point of view of the males, a Saturday night dance was a dismal failure if the schoolma'am failed to show up.

The era had its quota of characters. There was one I'll call Wesley Wayne (because that was not his name). Just the sound of his voice sent shivers of ecstasy up and down the spines of females for miles around. Staid married women, growing girls, and yes—even the sophisticated schoolma'ams came under his spell and worshipped at his shrine, but to no avail. The great Wesley was unreachable. A young bachelor, handsome as a Greek God, and the

1927. A road cutting crew.

1930's, The original road from Burns Lake to the Francois Lake ferry Landing. Many early settlers trod this road, now a paved highway. [Photo courtesy of Mrs. E. F. Rowland]

1928, near Rose Lake. Originally the Telegraph Trail, the "Gov't" road when this snap was taken, and now the Yellowhead Highway, this is the summit between the Bulkley Valley and the Lakes District. From here, water drains easterly to the Endako-Nechako-Fraser River and Westward to the Bulkley-Skeena system. [Photo courtesy of Mrs. E. F. Rowland.]

owner of a beautiful pinto saddle horse, (which he draped with hundreds of dollars worth of gaudy trappings), Wesley would arrive early at the Saturday night dance and take a seat next to the musicians. Fortunately for him, this was long before the advent of music amplification, or he would have been stone deaf at an early age.

He seldom left his post all night. He did not smoke or drink intoxicating liquor, which enhanced him all the more to his many female admirers, but unhappily for them he did not dance either. No, not our Wesley. He would sit for hours, dreamily tapping his toes in time to the music, politely but very firmly discouraging coy advances from females who never gave up hope that they would strike a responsive chord in the heart of the great one. But this was not to be—Wesley had a mission and nothing could distract him from doing his thing.

As the dance draws to its close he becomes more pre-occupied. He seems to be waiting and listening for something. Finally, the violin player rises to announce "take your partners for the home waltz". There is a lull as the musicians tactfully wait for sweethearts and husbands and wives to locate each other for the final dance of the evening, and Wesley's moment has finally arrived. With perfect timing, he slowly uncoils his magnificent length, faces the dancers and holds up his right hand. The crowd is hushed—although they know exactly what is coming—the charm and magnetism of Wesley is such that they stand transfixed. The silence and tension almost reaches the breaking point, and then the great one solemnly announces in a slow nasal twang "With your wife's cold feet in the middle of your back, there's no place like home". The spell is broken, the music breaks into the strains of "Home Sweet Home" and without a further word or glance in any direction, Wesley strides lithely and gracefully out to his patiently waiting steed and rides off into the night.

No professional performer ever took greater pride in his act, or made a better exit than did Wesley Wayne who performed his little ritual every Saturday night for many years. He is probably now in the stump farmers corner of Heaven, and no doubt on Saturday nights sits with the angel orchestra and waits patiently for the leader to announce the "Home Waltz".

14

The Dirty Thirties

By the end of the 1920's, the Bulkley Valley-Lakes District area was enjoying a period of slowly developing, quiet prosperity. Smithers, which had come into being as the railway's subdivisional headquarters was steadily growing into a major trading centre; Hazelton, its brief glorious years as the terminus of water transportation already passed into history, was struggling to establish a new identity as a mining town, and Burns Lake, trading centre for the vast, thinly populated "south country", had proudly erected a directional sign which read, BURNS LAKE - 1 MILE (pop 312). Houston and Telkwa were struggling to establish themselves but were still only flag stops. If one wanted to board

Decker Lake. The ultimate in degradation: Part of the 20¢ per day "slave camp" in the Depression days. [Photo courtesy of G. Saul]

Topley Circa 1939. The mechanical age arrives. This was among the first trucks used in the tie industry and was owned by Gordon Holmes of Topley. [Photo courtesy of Mr. Ivan Holmes.]

Moose were plentiful and calves were made into pets. This one is at Goodwins, Decker Lake. [Photo courtesy of Mrs. E. F. Rowland

Another pet moose at Alec Strimbolds', Topley. [Photo courtesy of Mrs. E. Strimbold]

the train, it was necessary to set out a flag by day or a lantern by night to get it to stop.

There were a few self supporting ranches, some sporadic mining activities, and three or four railway maintenance men working from each of the railway stations spaced at five to ten mile intervals along the line; but the bulk of the meager population and the mainstay of the economy were the stump farmer-broadaxe families scattered sparsely throughout an area about three miles each side of the railway, and along the shores of the large lakes south of Burns Lake.

After "toughing it out" for the first few years,

most of the homesteaders had compromised (Don't we all?) with their original dreams of becoming self supporting farmers and settled for a life style which involved making the main part of their living from working in the woods producing hewn ties and slowly developing their homesteads during the off season.

Historians usually refer to this decade as the Roaring Twenties, a period when North Americans made huge fortunes dealing in sky-rocketing stocks and bonds, generally raised hell and spent money like drunken sailors. The inhabitants of the Bulkley Valley and Lakes District must have been the squares of their day, for they were blissfully unaware of all these shenanigans as they doggedly labored at their self appointed task of developing a new frontier. When the news of the stockmarket crash of 1929 filtered in via the few battery operated radios scattered throughout the area, it meant nothing tangible—such events belonged only to a world they had left behind. However, as the poet John Donne so aptly points out, "No man is an Island".

During the summer of 1930, persistent rumours began to circulate that the railway was going to drastically curtail its purchases of hewn ties. These rumours were met with the same skepticism meted out to the man who predicts spring by March. There had always been such predictions—but spring had never arrived in March, and regardless of calamity howlers, the railway had always purchased all the ties made available. By the fall, however, rumour became reality when all sub-contractors were formally notified that only a very limited number, if any, of hewn ties would be accepted from them during the coming season—and the great Depression, later to be referred to in this area as the Dirty Thirties, had arrived.

The official Forest Service records of hewn ties cut within the area from Prince Rupert to Endako during the two decades of the stump-farm broadaxe era (1919-1938 inclusive) have kindly been made available to the writer and the figures present a graphic picture of how the Depression manifested itself in this remote part of the world.

During 1919, the railway purchased 277,000 hewn ties. The number bought then steadily increased each ensuing year, a total of 6,130,000 being loaded out during the 12 year period ending in 1930—an average of 510,000 per season. The output for the three years immediately previous to 1931 had reached record highs: 1928—719,000; 1929—868,000; 1930—702,000. (2,390,000 for the three year period.

In 1931, the purchases dropped to 274,000; 1932 to an unbelievable 76,000; 1933 to 91,000. These were the worst years but there was to be little real improvement until the latter part of the decade.

To summarize: During the initial 12 year period, 1919 to 1930 inclusive—total cut 6,130,000; Annual average 510,000. During the Dirty Thirties eight year period, 1931 to 1938 inclusive—total cut 1,660,000; Annual average 207,000. The total cut in the 20 year stumpfarm-broadaxe era—7,790,000.

It was a year or so before the full impact of the reduction of hewn tie purchases was felt. The sub-contractors kept on making a limited number of ties, and the contractor owned stores (and a few others) rather reluctantly, in most cases, extended credit for essentials in the expectation that the railway would eventually purchase the uncontracted-for ties. Such was not the case, however, and many ties thus "cut on spec" deteriorated on the landings and were eventually used as firewood or fenceposts.

The Depression soon manifested itself in other ways: Rail traffic was drastically curtailed, maintenance crews reduced and highway work restricted. Beef and farm produce could not be sold, except for ridiculously low prices. Beef was sold for as low as 4¢ a pound, and the ultimate low was reached in 1933 when the Provincial Government decided it could no longer afford to allocate money for such frivolous items as fighting forest fires. The very radical C.C.F. party was formed that same year and found many instant converts among the stumpfarmer-broadaxemen, who rationalized that any new political philosophy would be an improvement on one that blandly abandoned the country's natural resources and the rural population to the vagaries of forest fires.

It soon became obvious that the stores could not extend credit indefinitely and the humiliating "relief" system came into being. Relief, as it was administered during the Dirty Thirties, differed in one important aspect from Welfare as it is allocated today. Relief was not a handout. The recipients worked for what they received. Only in very extreme cases involving the sick or very old was there any deviation from this policy of "work before you get your handout".

It seems, in retrospect, that applying for work and duly carrying out the task allocated should not have been a degrading and humiliating experience.

Early 1930's, Telkwa. The early thirties was the "heyday" of the model T's. No traffic problems—you parked whichever way you were headed. [Photo courtesy of Mrs. W. H. Morris]

1936. Cheap transportation during the Dirty Thirties. The original Rose Lake school is shown [right centre] and in centre two moose [which were very plentiful during this period], are unconcernedly grazing.

1925. A typical tie-cutting "strip" in the Decker Lake area. At the end of the day the newly cut ties will be dragged into piles along the pre-cut sleigh-road. [Photo courtesy Mrs. E. F. Mould]

Len Ailport, 1922, immigrant from California on his "Mountain Meadow" ranch in the summit area near Rose Lake. The homesteads in this high elevation area were of the stump-farm variety and were abandoned in the Dirty Thirties.

Decker Lake, 1920's. Sam Moran and Julie Buer pose in front of newly built cabin. [Photo courtesy of Mrs. Buer]

1926. Thor Strimbold making ties south of Topley. He later had a distinguished career with the B.C. Forest Service and became Chief Protection Officer for the vast Prince Rupert Forest District. Now retired and living in Terrace, B.C. [Photo courtesy S. T. Strimbold]

However, it involved a "cap in hand" type of procedure that most applicants had travelled thousands of miles from their homelands and endured many hardships to avoid.

First, the applicant had to sign a form, swearing that he was absolutely destitute. This was followed by a series of interviews, during which a government official pried into every facet of the applicant's circumstances, almost to the extent of checking the food supplies, if any, on hand in the cupboards. Then, if and when the application was finally approved, the fledgling reliefer was instructed to report to a local foreman (usually a political heeler) to work out his "relief" on the main road, now Highway #16.

I am not sure how the scale of relief work allowances pertained to family sizes but I do know (I'll never forget) that as a married man with two children, during the period I was compelled to accept relief, I received $14.00 per month, our only income, for which I worked at hard manual labor for 35¢ an hour.

Initially there was not even this meager acknowledgment of the plight of the single unemployed, and as history now belatedly records, tens of thousands, including many from this area, roamed over the country begging for sustenance. Society conveniently eased its collective conscience by labelling the unfortunates hobos, bums and, as a clincher, that sure fire all inclusive catch phrase "reds and Commies".

The Federal government eventually was compelled to acknowledge that a problem existed in regard to the single unemployed and again history has recorded how the problem was met. Slave work camps were established throughout the country. I have searched for a milder term, but there just appears to be no other words to adequately describe a situation where citizens whose only misfortune was that the economy could not provide them with a job, were herded into tent camps, poorly fed, paid 20¢ A DAY for their labor, and kept under police surveillance. R.C.M.Police stool pigeons were infiltrated into the ranks to report on any "commie" who had guts enough to talk out against such indignities.

Ironically enough, one of these slave camps was established at Decker Lake, right in the heart of the best tie timber in what had heretofore been the habitat of the most fiercely independent people in the West. People were still there, but they were no longer "fiercely independent". The Dirty Thirties did not recognise such an idiosyncrasy.

The camp accommodated over 100 men, most of whom were shipped in from other areas of the province. There were eight men allocated to each 12' x 14' tent, and the only toilet facility was a single stall six holer. There were no recreation facilities and as the camp was situated in a low creek bed area, millions of mosquitoes added to the general misery. There was no fly repellent issued and the slaves spent most of their free hours huddled around smudges. They earned their meager food and 20¢ per day either shovelling gravel or pushing wheelbarrows in connection with local road construction and maintenance.

There were only a few local men forced into this camp. The single unemployed regional tie hackers mostly "shacked up", and some stores (most notably the S. Anderson Company) continued to extend credit to them for bare essentials.

The provincial government benevolently set up a system by which land taxes could be "worked out" (on road maintenance, usually gravel hauling at $6.40 per day for team, wagon and driver; $3.20 for laborer), but even so, many homesteads were abandoned and finally reverted for tax delinquency.

For those who toughed it out, existence was made a little less difficult by the influx of moose into the area. They had been sighted occasionally during the early 1920's, but the first real population explosion of these huge animals generally coincided with the start of the Great Depression, almost as if some Divine Power decided something should be done to counteract the stupidity and ineptness of the human authorities.

Moose were killed at all times of the year. In winter, the meat could usually be kept frozen by being hung out in the open, but exposure to the air caused it to turn black and tend to dry out. Many people solved this problem (and that of occasional winter thaws) by cutting the meat into meal size pieces and burying them in a big snow pile. When the housewife wanted a roast, she went out and dug for it. During the rest of the year, individuals would rotate in taking the responsibility of killing a moose and the meat would be shared among the families involved, who would can whatever could not be used while it was still fresh.

There were seasonal restrictions "on the books", but the authorities made no pretense at enforcing them. I suppose a government that was blandly allowing forest fires to go unchecked could hardly afford to jail homesteaders for killing moose to feed hungry kids. Actually there was no indiscriminate slaughter. There were few homesteaders in relation to the large moose population and the animals were killed for meat only, never for sport, so the limited amount of season killing did not slow down the rapid moose population increase to any appreciable extent.

Toward the end of the 1930's, the railway tie market became re-activated to some extent, but by then the mechanical age was beginning to make a belated appearance in the form of little portable sawmills which quickly began to replace the broadaxes.

Then the outbreak of War #2 in 1939 created an immediate unlimited demand for forest products and this brought almost instant prosperity to North Central B.C. It also triggered an inexorable march of progress, manifested by an influx of people, machines and big business into the district, which for better or worse spelled an abrupt "Finis" to the two decade stumpfarm-broadaxe era, and brings this little reminiscence to its end.

Glossary

Broadaxes: 7-8 lbs. Flat on one side. Used to hew off the sides of the trees after scoring.

Dryass: Heavy gunny sack periodically filled with dry hay. Used by tie haulers for sitting on.

God's Brothers: The railway's tie inspectors (Graders).

Jackpine: Name commonly applied to lodgepole pine from which most hewn ties were made. Actually jackpine does not grow naturally in central B.C. (But never try to convince an old time tie-maker of this).

Landings: Places where ties were piled along the railway sidings and spurs to await loading. Sites were allocated by Grand Trunk Pacific officials for a small fee.

Limeys: (Green Englishmen). Usually hard workers but an awfully long way from 'ome! They seemed determined to do everything the hard way.

Picaroon: Pick-like tool with a hooked end, used for handling hewn ties.

Pioneers: In the context of this book, those who established residence prior to the building of the Grand Trunk Pacific Railway (Pre 1914).

Prairie Chickens: Immigrants from the prairies. At the start most were very poor at bush work, but their general knowledge of ranching, farming and livestock was a valuable asset.

Remittance Men: Usually younger sons of well to do English families, receiving an allowance to remain in the colonies.

Roughlock: Short heavy chain that was threaded around rear sleigh runners when necessary to slow heavy loads coming down steep grades.

Scoring axes: 4½-5½ lbs. Usually double bitted. Used for "scoring" the timber down to within two inches of specified thickness.

Settlers: Generally the post World War One rail borne immigrants (1914-1924).

Spud: Tie peeling tool.

Snaabsaw: About three feet long without rakers; easy to set and sharpen. Well suited for cutting up the relatively small tie timber especially when frozen. (Some used standard crosscut saws).

Tiehacks: The makers of hewn ties. Most tie-makers disliked the term, as hacking was usually associated with poor axemanship; however once coined, the term stuck.

Turkey: The pack of a travelling tie-maker. The axe handles sticking out of the bedroll gave the appearance of a turkey.